YOUR KNOWLEDGE HAS VALUE

- We will publish your bachelor's and master's thesis, essays and papers

- Your own eBook and book - sold worldwide in all relevant shops

- Earn money with each sale

Upload your text at www.GRIN.com and publish for free

Bibliographic information published by the German National Library:

The German National Library lists this publication in the National Bibliography; detailed bibliographic data are available on the Internet at http://dnb.dnb.de .

Imprint:

Copyright © 2016 GRIN Verlag
Print and binding: Books on Demand GmbH, Norderstedt Germany
ISBN: 9783668775381

Kritish De

Studies on Land Use and Land Cover Patterns in Southern Dandakaryanaya with Special Reference to Forest Fragmentation in and around Bailadila Region

GRIN Verlag

CONTENTS

List of figures *

* all figures are the author's own work

CHAPTER I

An overview on Dandakaranya and objectives of the study

1.1 An overview on Dandakaranya

1.1.1 Dandakaranya : Etymology and mythology

In the ancient Hindu literatures Ramayana, Mahabharata and several Puranas two spiritually significant forests were mentioned namely Naimisharanya in the northern India and Dandakaranya in the south central India.

The word Dandakaranya is formed by joining the two separate words Dandaka and Aranya according to a rule of Sanskrit grammar. Aranya means a forest and Dandaka means punishment and Dandakaranya means Forest of Punishment.

It was described in Hindu mythology that Dandakaranya region was extended from Vindhya hill range, Narmada river and Mahanadi river in the north to Godavari river (in some mythology Krishna river) in the south, from Mahendragiri mountains of Eastern Ghat hills in the east to Wardha river in the west.

Near about the beginning of Treta yuga Dandaka was a country ruled by a king named Danda who was the youngest son of the legendary king Ikshvaku, son of Manu Vaivasvata and founder of the Solar Dynasty of kings. Ikshvaku, finding Danda a great fool and the most useless, banished him to this region because he was highly anxious of his actions. Ikshvaku got a capital city built for him from where Danda ruled. But Danda continued to lead a voluptuous life. Danda's kulaguru (royal guru) was Shukracharya who lived in an ashram located in the jungle surrounding Dandaka kingdom. Once, when Shukracharya was away, Danda visited the ashram and molested Shukracharya's daughter Araja, then left the ashram leaving Araja in trauma. When Shukracharya returned, Araja told the entire incident to him. This made Shukracharya very angry and he cursed Danda: *"In 7 days, you and your kingdom, all your people and army, shall die. For a hundred yojanas around your city, all life will be consumed by a rain of dust and death shall rule this sinner's kingdom."* Things happened as per the curse. All life was extinguished. Danda perished. Dandaka kingdom was laid waste; in consequence the kingdom became Dandakaranya - the forest of punishment, a region of dense wild forest through which even sunlight did not pass.

Later, Dandakaranya became part of colonial state of Lanka under the reign of Ravana. Khara, a man-eating rakshasa (demon) and younger brother of Ravana was governor of the Dandakaranya province. Dandakaranya became a stronghold of the Rakshasa (demon) and then Dandakaranya was called the forest of demons.

In the epic Ramayana, many of the events described in Aranya Kanda were happed in Dandakaranya.

1.1.2 Dandakaranya : Present extents

In modern India, the Dandakaranya region does not have any political or administrative boundary. But districts of Bastar division (which includes 7 districts namely Bastar, Bijapur, Dantewada, Kanker, Kondagaon, Narayanpur and Sukma) of Chattishgarh state; 4 districts (namely Kalahandi, Koraput, Malkangiri and Nabarangpur) of Odisha state; Gadchiroli and Gondia districts of Maharastra state and Khambam district of Telengana are considered as present extends of Dandakaranya.

1.1.3 Dandakaranya: Gelogy

Geologically the Dandakaranya region is present on the the Bastar craton. This Bastar craton is rectangular in shape, bounded by Godavari graben in its southwest, Eastern Ghat Mobile Belt in the southeast and in the northwest the Sausar Mobile Belt (Satpura Orogeny) and by Mahanadi graben in the northeast (Sridhar et al, 2015). The Bastar craton consists of widespread Tonalite - Trondhjemite - Granodiorite (TTG) gneisses, granite gneiss and migmatites namely

Amgaon Group, supracrustals of Bengpal Group, Sukma Group, Sakoli fold belt and Bhopalpatanam granulites occur as enclaves within it (Sridhar et al, 2015). The younger supracrustals in the Bastar craton include the Bailadilla Group, Dongargarh Supergroup and the Sonakhan Group. These supracrustals are overlain by platformal sediments of Chhattisgarh, Indravati, Sukma, Abhujmar, Karihar, Kaskal, Ampani and Phakals basins (Sridhar et al., 2015).

1.1.4 Dandakaranya : Minerals and mines

The Dandakaranya region is a rich source of minerals and fossil fuel. High-grade aluminium ore (bauxite), chlorite, coal, corundum, gemstone (acquamarine, alexandrite, beryle, cat's eye, garnet, moonstone, ruby, sapphire, topaz and tourmaline), granite (coloured decorative stone), graphite, iron ores (Hematite and Titano-Magnetite), limestone and dolomite, manganese, mica, quartz, tin ore (cassiterite) are mined from this region. Some of the largest mining areas of the world such as Bailadila iron ore range and Koraput bauxite ore range are present in this area.

1.1.5 Dandakaranya : Forest and wildlife

Forests of Dandakaranya are of tropical moist deciduous forest (dominants mainly deciduous but sub-dominants and lower story largely evergreen top canopy even and dense but 25 m high) and tropical dry deciduous forest (entirely deciduous or nearly so top canopy uneven rarely over 25 m high) type. Sal (*Shorea robusta*) is the dominant plant in Dandakaranya in association with Mahua (*Madhuca longifolia*), Amla (*Phyllanthus emblica*), Bija (*Pterocarpus marsupium*), Dhawra (*Anogeissus latifolia*), Karra (*Cleistanthus collinus*), Karunj (*Pongamia pinnata*), Kusum (*Schleichera oleosa*), Saja (*Terminalia tomentosa*), Tendu (*Diospyros melanoxylon*), Bamboo (*Dendrocalamus strictus*) and Fishtail palm / Sulphi (*Caryota urens*). Beside there are also rich patches of grasslands present in Dandakaranya. The floristic elements show some affinities with the flora of Assam- Bihar, Eastern Ghats, Western Ghats as well as Central Indian region (Govekar, 2008).

In the present extends of Dandakaranya, two National Parks (Indravati National Park and Kanger Valley National Park) and five Wildlife Sanctuaries (Balimela Wildlife Sanctuary, Barnawapara Wildlife Sanctuary, Karlapat Wildlife Sanctuary, Pamed Wildlife Sanctuary and Sitanadi Wildlife Sanctuary) are established where natural and wild fauna and flora are protected by law. Besides, there are thousands of sacred groves where wild flora and fauna are protected traditionally by the aborigines. In 2007, Ministry of Environment, Forests and Climate Change, Government of India designated hilly forest area of Abujmarh - the unknown hills, spread over about 3,900 square km as a potential sites to be designated as Biosphere Reserve.

Some important wild fauna of Dandakaranya are Royal bengal tiger (*Panthera tigris tigris*), Indian leopard (*Panthera pardus fusca*), Indian elephant (*Elephas maximus indicus*), Black buck (*Antilope cervicapra*), Wild dog (*Cuon alpinus*), Sloth bear (*Melursus ursinus*), Barking deer (*Muntiacus muntjak*), Sambar (*Rusa unicolor*) etc.

Some of the fauna are endemic to Dandakaranya i.e. they do not found any other place of the world. For example, a troglobitic (lives entirely in the dark parts of caves) population of *Indoreonectes evezardi*, Day, 1872 commonly called stone loache, a fish under family Nemacheilidae is found only in Gopansar cave (Kotumsar cave) of Kanger Valley National Park.

1.1.6 Dandakaranya : The tribals

The tribals of Dandakaranya region are unique and have distinctive tribal culture and heritage. Each tribal group enjoys their own unique traditional living styles. Each tribe has developed its own dialects and differs from each other in their costume, eating habits, traditions and even worships different form of god and goddess. Some important tribal groups of Dandakaranya are Abhuj Maria, Bison Horn Maria, Bhatra Bhatra, Bonda, Dhurvaa, Gond, Halbaa, Kondha, Muria, Paraja and Shabar.

1.1.7 Dandakaranya : The DNK project

In spite of spiritual and religious significance, the Dandakaranya region was in ignorance till 1958, when Government of India started The Dandakaranya Project (DNK Project). This DNK Project was started to settle the Bengali Hindu refugee came from Bangladesh (then East Pakistan) after independence and separation (in 1947) of nations. During DNK projects and other projects some development activities such as new railway tracks, highways and power plants were build up in this region.

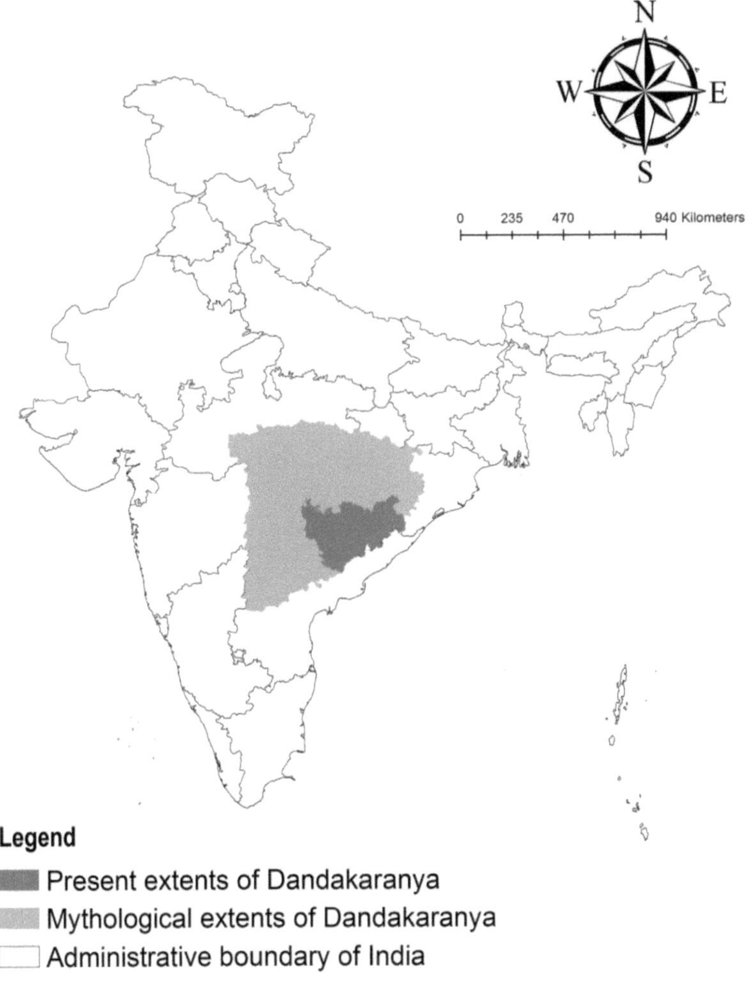

Plate 1: Present and mythological extents of Dandakaranya

1.1.8 Dandakaranya : BRGF program

The Dandakaranya region is suffering from the greatest illiteracy and poverty in India. A total of 22 districts of which 15 districts (Bastar, Bijapur, Bilaspur, Dantewada, Dhamtari, Jashpur, Kabirdham, Kanker, Korba, Koriya, Mahasammund, Narayanpur, Raigarh, Rajnandgaon and Sarguja) from Chattishgarh state, 4 districts (Kalahandi, Koraput, Malkangiri and Nabarangpur) from Odisha, 1 district (Khammam) from Telengana and 2 districts (Gadchiroli and Gondia) from Maharastra which come under present extends of Dandakaranya region are included in Backward Regions Grant Fund (BRGF) program of Government of India.

1.2 Objectives of the study

To know changes of land use land cover pattern in southern Dandakaranya.
To know forest fragmentation in and around Bailadila hill iron ore range.

CHAPTER II

Land use land cover changes of southern Dandakaranya

2.1 Introduction

Natural Resources Management and Environment Department of Food and Agriculture Organization of the United Nations (2000) defines land cover as the observed (bio)physical cover on the earth's surface and land use as the arrangements, activities and inputs people undertake in a certain land cover type to produce, change or maintain it. Simply, land cover is the physical material at the surface of the earth and land use is the human modification of natural environment or wilderness into built environment such as fields, pastures and settlements (Samanta and Hazra, 2012). Land use is never static, but it is constantly changing in response to dynamic interaction between drivers and feedback from land-use change to these drivers (Lambin, Geist and Lepers, 2003). Land-use change is a spatial property observed at the scale of a landscape. It is the sum of many small, local-scale changes in land allocation that reinforce or cancel each other (Lambin, Geist and Lepers, 2003).

Concerns about land-use/cover change emerged in the research agenda on global environmental change several decades ago with the realization that land surface processes influence climate (Lambin, Geist and Lepers, 2003). Over the last few decades, numerous researchers have improved measurements of land-cover change, the understanding of the causes of land-use change, and predictive models of land-use/cover change (Lambin, Geist and Lepers, 2003). Knowledge of land use and land cover change is important for many perspective planning and natural resource management initiative (Samanta and Hazra, 2012).

In present times, the remote sensing and GIS technologies are widely used to study the characteristics and changes of land use and land cover. Change detection analysis using multitemporal satellite data attempts to discriminate areas of land cover change between dates of imaging (Samanta and Hazra, 2012).

Till date, very few remote sensing and GIS based works on land-use/cover as well as other aspects in the Dandakaranya, especially southern Dandakaranya have done.In 1998, a study on forest cover of the area was conducted by a committee constituted by Ministry of Environment, Forest and Climate Change consisting of representative from Forest Survey of India (FSI), Botanical Survey of India (BSI), Indian Bureau of Mines (IBM), Geological Survey of India (GSI), National Remote Sensing Agency (NRSA), Indian School of Mines (ISM), Federation of Indian Mining Industries (FIMI) and Steel Authority of India (SAIL) (Central Pollution Control Board, 2007). Reddy et al. (2009) assessed large-scale deforestation of Nawarangpur district, Orissa. Pattanaik et al. (2011) had studied deforestation in Malkangiri district, Odisha. Thakur et al. (2014) characterized the land use, vegetation structure, and diversity in the Barnowpara Sanctuary. Kumari (2015) had studied on management of water runoff in in Bailadila Range using geoinformatic techniques. Sudhakar, Pujar and Anupama (2015) studied the palaeovegetation and reconstruction of vegetation types in Jagdalpur forest division using remote sensing and GIS techniques.

In the present study, land use and land cover change of southern Dandakaranya over the span of 25 years (from 1991 to 2016) is documented for the first time.

2.2 Materials and methods

2.2.1 Study area

Present study area is located in the southern part of Dandakaranya region. It is bounded by Indravati river on the north and north - west, Kolab (Sabai) river on the east, Godavari river on the south and south west. It includes 4 districts of the state of Chattishgarh, namely Bastar, Bijapur, Dantewada and Sukma. Total area of the study site is about 22097.55 square km.

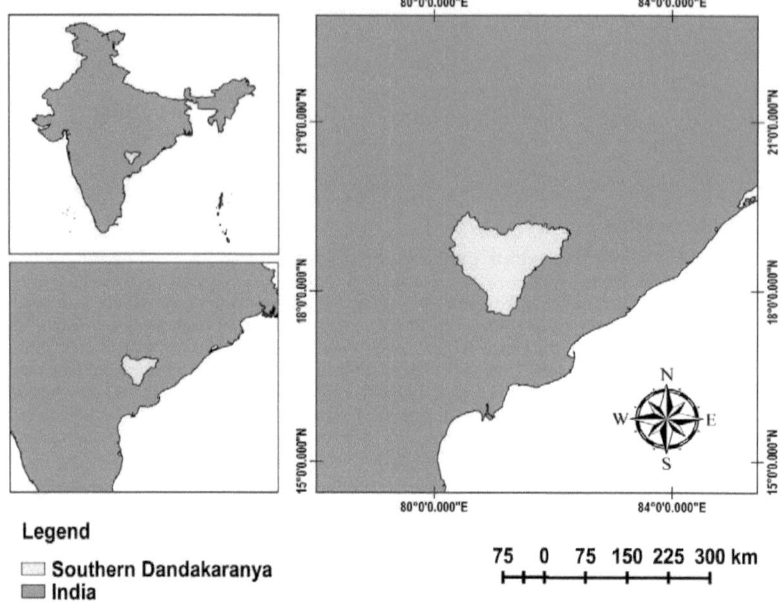

Plate 2: Location of southern Dandakaranya

2.2.2 Methodology for land use land cover change detection

Landsat satellite images of three years - 1991, 2004 and 2016 was collected from USGS Glovis. The collected images were subjected to several processing techniques namely layer stack, spectral enhancement and mosaicking. The area of interest (in present case, southern Dandakaranya) was digitized from Google Earth. The area of interest (in present case, southern Dandakaranya) was extracted from the processed raster file. The images were subjected to visual interpretation and supervised classification for analysis of land use land cover changes. Land use land cover was classified into 5 class namely dense forest, moderately dense forest, open forest, unused land and agriculture and settlement. For analysing the change rate for land use land cover assessment, following equation proposed by Puyravaud (2003) was used:

Change rate = [{1/(t2-t1)} * ln(A2/A1)]

> Where,
> > t1 = base year
> > t2 = current year
> > A1 = area of a class in base year
> > A2 = area of a class in current year
> > ln = natural logarithm

NDVI images of the study area were generated by following equation:

NDVI = (NIR band - Red band) / (NIR band + Red band)

For all geoinformatic and statistical procedures Erdas Imagine 2011, Arc GIS 10.3, QGIS 2.8.8, Microsoft Office Excel 2007 and R 3.1.1 software were used.

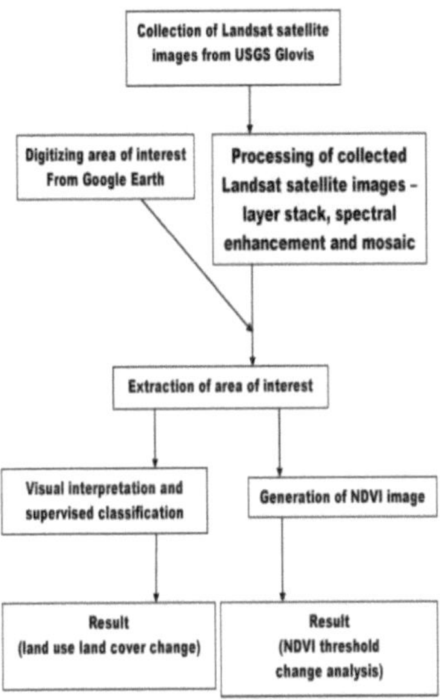

Figure 1: Conceptual frame work of methodology to study land use and land cover changes of southern Dandakaranya

2.3 Results

2.3.1 Land use land cover changes

From the study it was found that dense forest cover was decreased from 5054.521 square km (22.9% of total area) in 1991 to 4744.297 square km (21.5% of total area) in 2004 to 2707.677 square km (12.3% of total area) in 2016, moderately dense forest cover was increased from 5133.42 square km (23.2% of total area) in 1991 to 6473.01 square km (29.3% of total area) in 2004, then decreased to 6350.88 square km (28.7% of total area) in 2016; open forest cover was decreased from 5251.96 square km (23.8% of total area) in 1991 to 4052.237 square km (18.3% of total area) in 2004, then increased to 4559.491 square km (20.6% of total area) in 2016; unused land was decreased from 2547.58 square km (11.50% of total area) in 1991 to 2300.19 square km (10.40% of total area) in 2004, then increased to 3104.99 square km (14.10% of total area) in 2016; agriculture and settlement area was increased from 3806.3 square km (17.20% of total area) in 1991 to 4209.908 square km (19.10% of total area) in 2004, then again increased to 5374.511 square km (24.30% of total area) in 2016.

Table 1: Comparative account of land use land cover in southern Dandakaranya in 3 years - 1991, 2004 and 2016

	1991		2004		2016	
	Area	%	Area	%	Area	%
Unclassified	303.768	1.40%	317.907	1.40%	-	-
Dense forest	5054.521	22.90%	4744.297	21.50%	2707.677	12.30%
Moderately dense forest	5133.42	23.20%	6473.01	29.30%	6350.88 2	628.70%
Open forest	5251.96	23.80%	4052.237	18.30%	4559.491	20.60%
Agriculture and settlement	3806.3	17.20%	4209.908	19.10%	5374.511	24.30%
Unused land	2547.58	11.50%	2300.19	10.40%	3104.99	14.10%

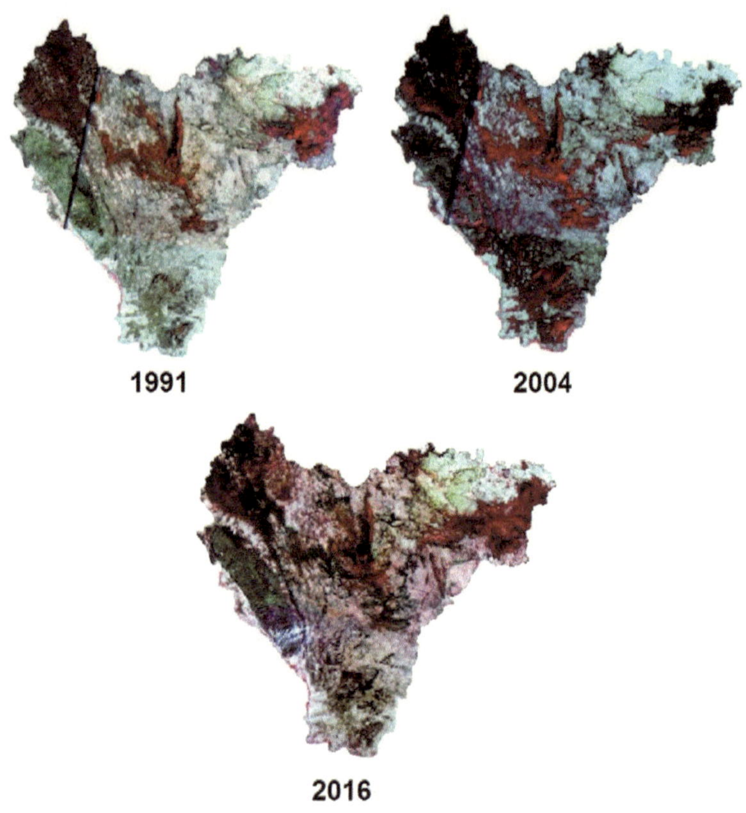

Plate 3: Landsat image of southern Dandakaranya in 3 years - 1991, 2004 and 2016

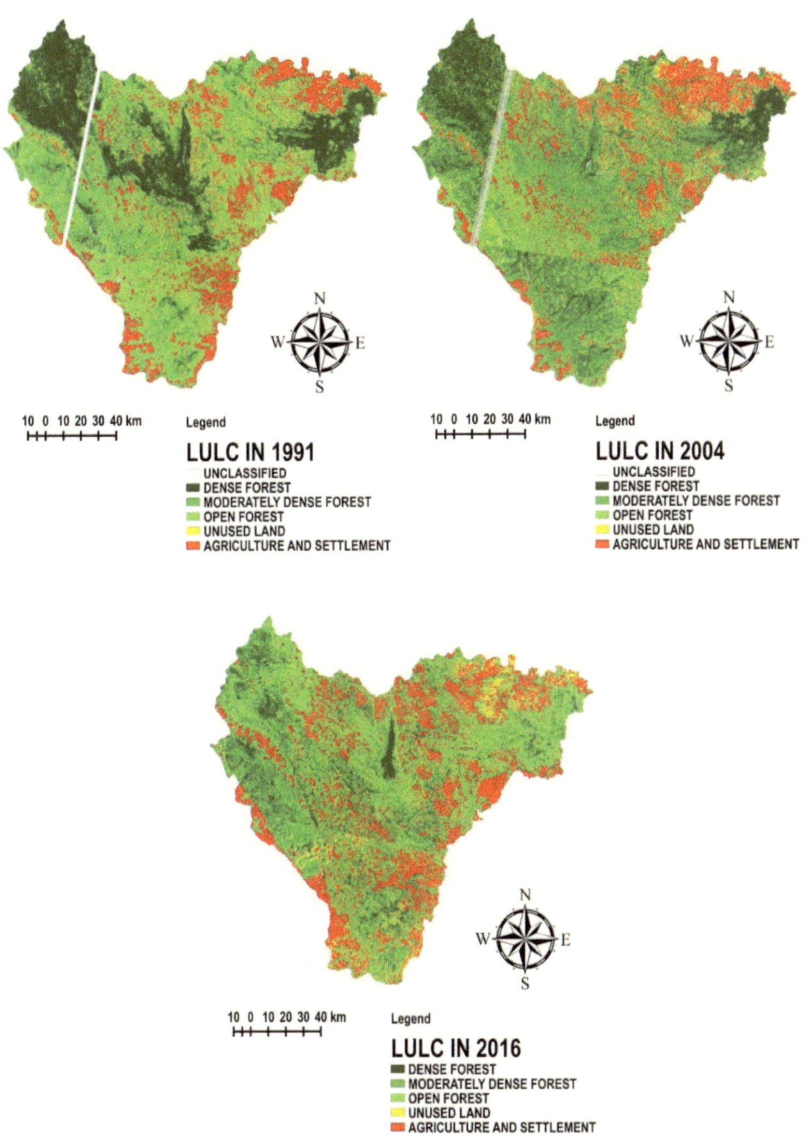

Plate 4: Land use land cover of southern Dandakaranya in 3 years - 1991, 2004 and 2016

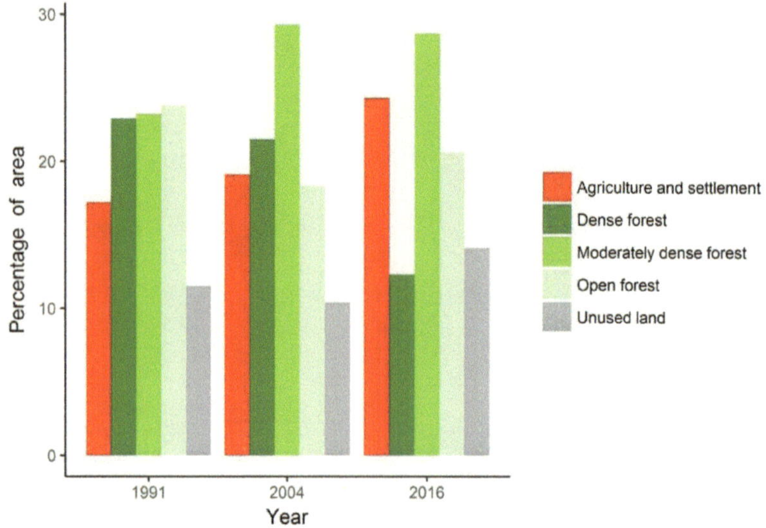

Figure 2: Comparative account of land use land cover in southern Dandakaranya in 3 years - 1991, 2004 and 2016

2.3.2 Land use land cover change rate

From the study it was found that change rate of dense forest was negative for all the time duration. From 1991 to 2004, change rate of dense forest was ↓0.005, from 2004 to 2016 change rate of dense forest was ↓0.047 and as a whole, from 1991 to 2016, change rate of dense forest was ↓0.025 (↓ indicates decreasing). From 1991 to 2004, change rate of moderately dense forest was 0.018, from 2004 to 2016 change rate of moderately dense forest was ↓0.002 and as a whole, from 1991 to 2016, change rate of moderately dense forest was0.009. From 1991 to 2004, change rate of open forest was ↓0.02, from 2004 to 2016 change rate of open forest was 0.01 and as a whole, from 1991 to 2016, change rate of open forest was ↓0.006. From 1991 to 2004, change rate of unused land was ↓0.008, from 2004 to 2016 change rate of unused land was 0.025 and as a whole, from 1991 to 2016, change rate of unused land was 0.008. From the study it was found that change rate of agriculture and settlement was positive for all the time duration. From 1991 to 2004, change rate of agriculture and settlement was 0.008, from 2004 to 2016 change rate of agriculture and settlement was 0.02 and as a whole, from 1991 to 2016, change rate of agriculture and settlement was 0.014.

Table 2: Land use land cover change rate assessment in southern Dandakaranya (↓indicates decreasing rate)

Change rate of	Duration		
	1991 to 2004	2004 to 2016	1991 to 2016
Dense forest	↓0.0049	↓0.0467	↓0.0250
Moderately dense forest	0.0178	↓0.0015	0.0085
Open forest	↓0.0199	0.0098	↓0.0057
Agriculture and settlement	↓0.0079	0.0250	0.0079
Unused land	0.0076	0.0204	0.0138

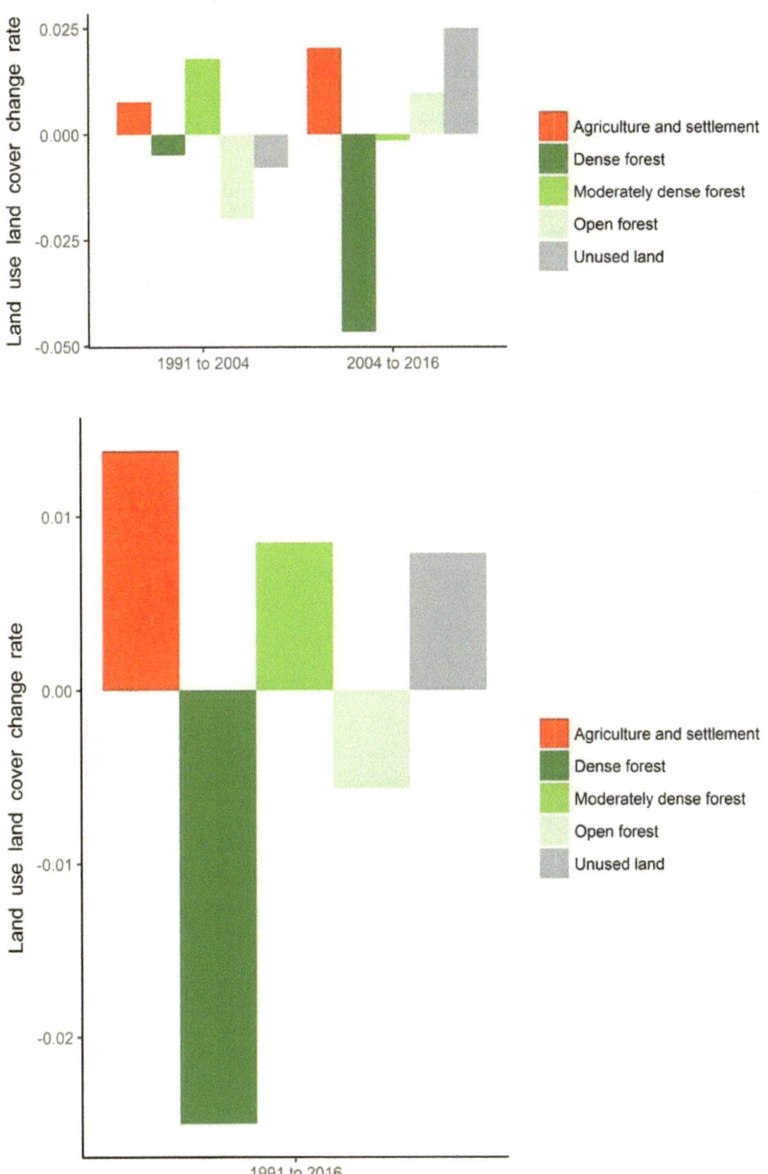

Figure 3: Comparative account of land use land cover change rate assessment in southern Dandakaranya

2.4 Discussion

Land cover change is a dynamic process taking place on the bio-physical surfaces that have taken place over a period of time and space is of enormous importance in natural resource studies (Jayanth, Kumar, Koliwad and Krishnashastry, 2016). Land-use change can be understood using the concepts of complex adaptive systems and transitions (Lambin, Geist and Lepers, 2003). Integrated, place-based research on land-use/land-cover change requires a combination of the agent-based systems and narrative perspectives of understanding (Lambin, Geist and Lepers, 2003). Proper planning, management, and monitoring of the natural resources depend on the availability of accurate land use information (Thakur, Swamy and Nain, 2014). Present study provides information about changes of land use and land cover as well as NDVI thresholds in southern Dandakaranya region over the period of 25 years (from 1991 to 2016). It is recommended that further time series studies should be carried out with high resolution satellite/aerial data to investigate in detail the changes of land use land cover, vegetation distribution and characteristics, forest cover fragmentation in the southern Dandakaranya region. These types of research will help the government and NGOs to formulate and imply necessary agricultural, environmental a forest management strategies for betterment of livelihood of people and ecosystem of Dandakaranya.

CHAPTER III

Forest fragmentation in and around Bailadila hill range

3.1 Introduction

Landscape ecology is the study of structure, function, and change in a heterogeneous landscape composed of interacting ecosystems (Smith and Smith, 2012). In a landscape, one or more area of habitat that differs from its surroundings and has sufficient resources to allow a population to persist which are called patches (Smith and Smith, 2012). If a landscape has more than one patch, then these patches may be connected with each other with corridors - a strip of a particular type of vegetation that differs from the land on both sides but similar with the connecting patches. At the outer boundary of habitat patches, a narrow sharp zone of habitat transition present, which is called edge. However, the habitat edge is not a discrete boundary line around a patch, it is a fuzzy three-dimensional zone that straddles both sides of the patch-matrix boundary, and the intensity of edge influence may be variable and asymmetrical around the physical vegetation boundary (Didham, 2010). Habitat patches and corridors play important role in dynamics of metapopulation (a population broken into sets of subpopulations held together by dispersal or movements of individuals among them, Smith and Smith, 2012). In the context of metapopulation and habitat patches, some patches act as source patches or source habitat (area of habitat in which a subpopulation of a species produces more individuals than needed for self maintenance, thus contributing to emigration, Smith and Smith, 2012) and some patches act as sink patches or sink habitat (a habitat area that receives immigrants from a source habitat, but in which the subpopulation would continually decrease in size because of mortality and poor reproductive success without continual immigration from excess individuals in a source habitat, Smith and Smith, 2012).

Habitat fragmentation is an umbrella term describing the complete process by which habitat loss results in the division of large, continuous habitats into a greater number of smaller patches of lower total area, isolated from each other by a matrix of dissimilar habitats, and is not just the pattern of spatial arrangement of remaining habitat (Didham, 2010). Habitat fragmentation is considered as one of the main threats to biodiversity, causing land covers or habitats to break to smaller and less connected pieces and thus reducing their connectivity (Zebardast et al., 2011). Landscape fragmentation caused by roads, railway lines, extension of settlement areas, etc. enhances the dispersion of pollutants and acoustic emissions and affects local climatic conditions, water balance, scenery, and land use (Jaeger, 2000). Habitat loss and associated species loss are primarily a result of the acceleration of land-use changes (Hilty, Lidicker and Merenlender, 2006). Therefore, it is important to study land-use change as the root cause of the biodiversity crises (Hilty, Lidicker and Merenlender, 2006).

In the Bailadila hill range of southern Dandakaranya region high quality grade (+65% to +66% Fe) iron ore hematite is deposited which is mined by National Mineral Development Corporation, a Navaratna Public Sector Enterprise under Ministry of Steel, Government of India at Kirandul mine complex and Bacheli mine complex. In Bailadila hill range. Iron ore are extracted by open cast method, which has huge negative impact on environment such as deforestation, air pollution, soil pollution and noise pollution. Indian Bureau of Mines (2015) reported that in Bailadila sector, the deforestation is taking place due to mining and waste dumping.

Till date, very few studies are done on environmental and ecological issues of Bailadila hill range of southern Dandakaranya region using geoinformatic technologies. In 1998, a study on forest cover of the area was conducted by a committee constituted by Ministry of Environment, Forest and Climate Change consisting of representative from Forest Survey of India (FSI), Botanical Survey of India (BSI), Indian Bureau of Mines (IBM), Geological Survey of India (GSI), National Remote Sensing Agency (NRSA), Indian School of Mines (ISM), Federation of Indian Mining Industries (FIMI) and Steel Authority of India (SAIL) (Central Pollution Control Board, 2007). Kumari (2015) studied runoff water management in Bailadila range, Kirandul.

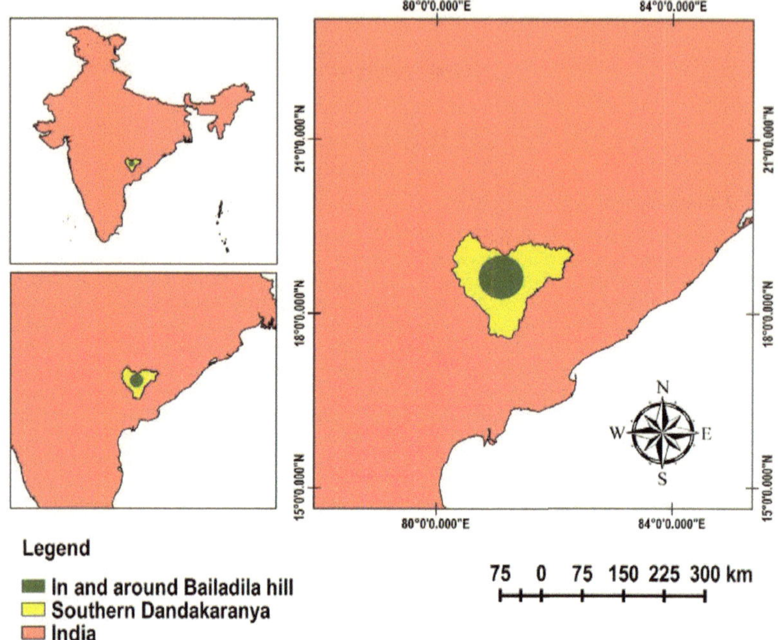

Legend

■ In and around Bailadila hill
□ Southern Dandakaranya
■ India

Plate 5: Location of the Bailadila hill region in southern Dandakaranya

3.2 Materials and methods

3.2.1 Study area

- Etymology: The term Bailadila comes from two words viz. baila means ox and dila means hump. As the Bailadila hill range looks like hump of an ox, it is so named by the native people.

- Location: The Bailadila hill range is located between between 18.668333°N to 18.69583°N latitudes and 81.18333°E to 81.20639°E longitudes at the district of Dantewada of the state of Chattisgarh. The Bailadila hill range is about 36 km long and 10 km wide (Kumari, 2015).

- Geology :Geo-morphologically, the terrain of the Bailadila is characterized by relict hill ridges with cliffs due to hard resistant ore body or iron formations, terraces formed by lateritisation at elevations of around 1000 to 1100 metre above MSL and deflected profile due to the above (Central Pollution Control Board, 2007). The highest peak of the area is about 1276 m above MSL and the entire range approximately forms a Y shape, with the tip pointing north direction (Central Pollution Control Board, 2007). The lower undulating plains of elevation varying from 300 meters to 400 metre have occasionally hills rising up to 600 Metres above MSL (Central Pollution Control Board, 2007).

- Climate: Average annual rainfall of 2660mm to 3000mm (Central Pollution Control Board, 2007). Low cloud commonly shrouds the mountain tops during July and August (Central Pollution Control Board, 2007). The temperature of the region is generally moderate with annual day average recording about 24°C to 35°C and the night average about 11°C to 17°C (Central Pollution Control Board, 2007). During summer seasons, the climate turns slightly arid with Relative humidity dipping down to about 20% and the temperature rising to about 40°C (Central Pollution Control Board, 2007). The predominant wind velocity is ranging from 19 to 29 kmph with SW and NE directions (Central Pollution Control Board, 2007). During monsoon and pre-monsoon seasons, the wind velocities touch as high as 60-70 kmph (Central Pollution Control Board, 2007)

- Hydrology: The entire region is a part of the Godavari river basin (Central Pollution Control Board, 2007). There are a number of perennial streams flowing from the hills (Central Pollution Control Board, 2007). The eastern slope of the area drains through streams, which flow towards north east to Sankani river (Central Pollution Control Board, 2007). Drainage in between the eastern and western ridges is through two streams flowing in opposite direction, i.e. Galli nalla towards south and Sankani nalla towards north (Central Pollution Control Board, 2007). Sankani nalla cuts across the eastern ridge near Jhikra village and flows down east and north-east and becomes Dantewada river which ultimately flows west and joins Indravati river (Central Pollution Control Board, 2007). The western slope of the area is drained by Mari nadi, Berudi nadi and other small streams, all of which meet river Indravati at different points (Central Pollution Control Board, 2007). Southern part of the area is drained through Malinger nadi joining Sabori river and Galli nalla to Talperu river, all again flow in to Godavari river (Central Pollution Control Board, 2007).

- Vegetation: It harbours rich biodiversity. Its upper ridges are covered with dense deciduous forest and the lower slopes and banks of the nalas consist of semi evergreen plant species (Kumari, 2015). Varied physiographic elements and microclimatic conditions, in association with vegetation, have made this area a treasure house of medicinal plants (Kumari, 2015).

- Mining: The Bailadila hill range, where high quality grade (+65% to +66% Fe) iron ore hematite is deposited if fall under Zone - B (Central Zone) of Iron Ore Zones in India (Central Pollution Control Board, 2007). The iron ore is mined by National Mineral Development Corporation. The Bailadila iron ore range is divided into 14 deposits (Central Pollution Control Board, 2007). Deposits 1 to 5 occur in the western range, deposits 6 to 12 occur in the eastern ridge, while deposit 13 and deposit 14 occur in the southern closure of the ridges (Central Pollution Control Board, 2007).

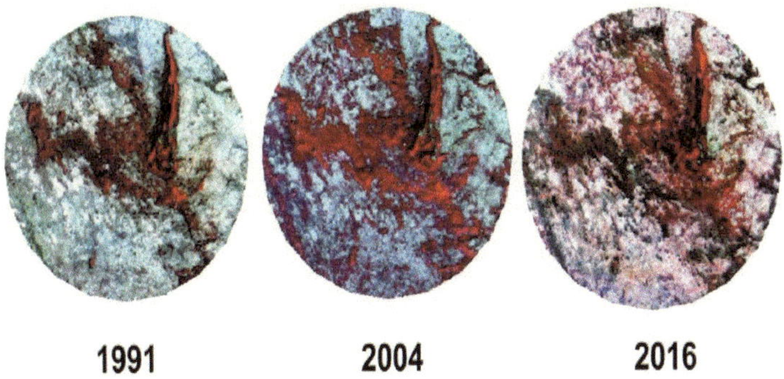

Plate 6: Landsat satellite image of Bailadila hill and surroundings of 3 years - 1991, 2004 and 2016

3.2.2 Methodology

3.2.2.1 Image processing

Landsat satellite images of three years - 1991, 2004 and 2016 was collected from USGS Glovis. The collected images were subjected to several processing techniques namely layer stack, spectral enhancement and mosaicking. The area of interest (in present case, area around Bails-dila hill range) was digitized from Google Earth. The area of interest (in present case, area around Bailsdila hill range) was extracted from the processed raster file.

The extracted raster file was reprojected from 30m * 30m cell size to 1 km * 1 km cell size.

The reprojected raster was subjected to supervised classification and visual interpretation.

Analysis of changes and fragmentation of forest cover was performed on the classified raster image.

For all geoinformatic and statistical procedures Erdas Imagine 2011, Arc GIS 10.2.2, QGIS 2.6.1, Microsoft Office Excel 2007, R 3.1.1 and PAST 3.05 software were used.

Forest was classified into 3 classes namely dense forest area, moderately dense forest area and open forest area.

3.2.2.2 Analysis of forest cover change rate

For analysing the change rate for deforestation assessment, following equation proposed by Puyravaud (2003) was used:

Change rate = [{1/(t2-t1)}* ln(A2/A1)]

        Where,

                t1 = base year

                t2 = current year

                A1 = area of a class in base year

                A2 = area of a class in current year

                ln = natural logarithm

3.2.2.3 Analysis of forest fragmentation

Total ten parameters generated by LecoS QGIS python plugin (Jung, 2013) namely Area, Landscape proportion, Number of patch, Largest patch size, Mean patch area, Patch density, Largest patch index, Edge length, Edge density and Landscape shape index were used to asses forest fragmentation of the area. A brief description of these landscape matrix parameters are given bellow -

- Area: It is the sum of the area of all the patches of a particular class.

- Landscape proportion : It represents proportion of area of a particular class to the total landscape area. Landscape proportion = (Area of a class)/(Total landscape area)

- Number of patch: Patch is an area of habitat that differs from its surroundings and has sufficient resources to allow a population to persist (Smith and Smith, 2012). Only information about number of patches provide no information about landscape matrices, but if total landscape area of interest held constant, then it acts as one of the most valuable basis for computation of other landscape matrix parameters.

- Largest patch size: It represents the area of the largest patch of a particular class in the landscape. It is used to calculate the largest patch index.

- Mean patch area: It is the ratio of area of a class to the numbers of the patches present in that class. Mean patch area = (Area of a class)/(Numbers of patch in that class)

- Patch density: It is the ratio of numbers of patches of a class to the total landscape area. Patch density = (Numbers of patch)/(Total landscape area)

- Largest patch index: Largest patch index at the class level quantifies the percentage of total landscape area comprised by the largest patch. As such, it is a simple measure of dominance (McGarigal, Cushman and Ene, 2012). LPI approaches 0 when the largest patch of the corresponding patch type is increasingly small. LPI = 100 when the entire landscape consists of a single patch of the corresponding patch type; that is, when the largest patch comprises 100% of the landscape (McGarigal, Cushman and Ene, 2012). Largest Patch Index = {(Greatest patch area)/(Total landscape area)} * 100

- Edge length: It is the sum of edge length of a particular class.

- Edge density: Edge density at the class level reports edge length on a per unit area basis that facilitates comparison among landscapes of varying size (McGarigal, Cushman and Ene, 2012). Edge density = (Edge length)/(Total landscape area)

- Landscape shape index: It is 1.00 when the landscape contains a single square shaped patch and increases without limit as the landscape shape become more irregular and/or length of the edge increases (Howard, 2005). Landscape shape Index = (0.25 *Total length of edge)/(√Total landscape area)

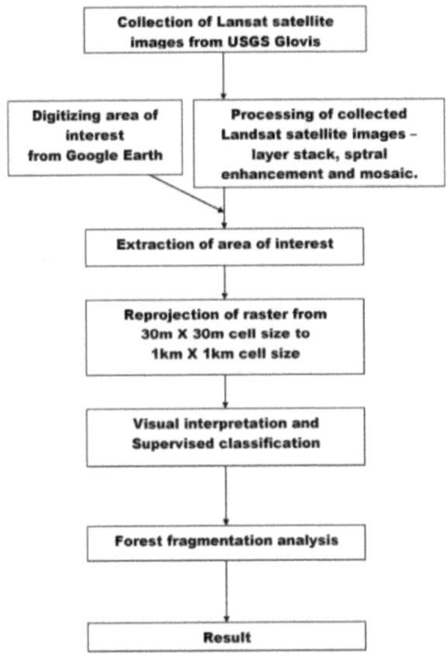

Figure 4: Conceptual frame work of methodology to study forest fragmentation in and around Bailadila hill range

3.3 RESULTS

3.3.1 Forest cover change

From the study it was found that total forest cover area is decreasing. In 1991, a total of 3300 square km area (63.425% of total area) was covered by forest, while in 2004 it was 3224 square km (61.964% of total area) and in 2016 it is 3207 square km (61.638% of total area). The rate of forest cover change between 1991 to 2004 was ↓0.002 and between 2004 to 2016 was ↓0.0004. Overall rate of forest cover change between 1991 to 2016 was ↓0.001 (↓ indicates decreasing).

Table 3: Forest cover change rate assessment in and around Bailadila hill (↓ indicates decreasing rate)

	1991 - 2004	2004 - 2016	1991 - 2016
Total forest cover area	↓0.0018	↓0.0004	↓0.0011
Dense forest area	↓0.0076	↓0.0919	↓0.0481
Moderately dense forest area	0.0184	↓0.0099	0.0048
Open forest area	↓0.0245	0.0353	0.0042

Dense forest cover was decreased from 592 square km in 1991 to 536 square km in 2004 (with a change rate of ↓0.008 between 1991 to 2004) to only 178 square km in 2016 (with a change rate of ↓0.092 between 2004 to 2016).

Moderately dense forest cover was increased from 1325 square km in 1991 to 1682 square km in 2004 (with a change rate of 0.018 between 1991 to 2004), then it was decreased to 1493 square km in 2016 (with a change rate of ↓0.01 between 2004 to 2016).

Open forest cover was decreased from 1383 square km in 1991 to 1006 square km in 2004 (with a change rate of ↓0.025 between 1991 to 2004), then it was increased to 1536 square km in 2016 (with a change rate of 0.035 between 2004 to 2016).

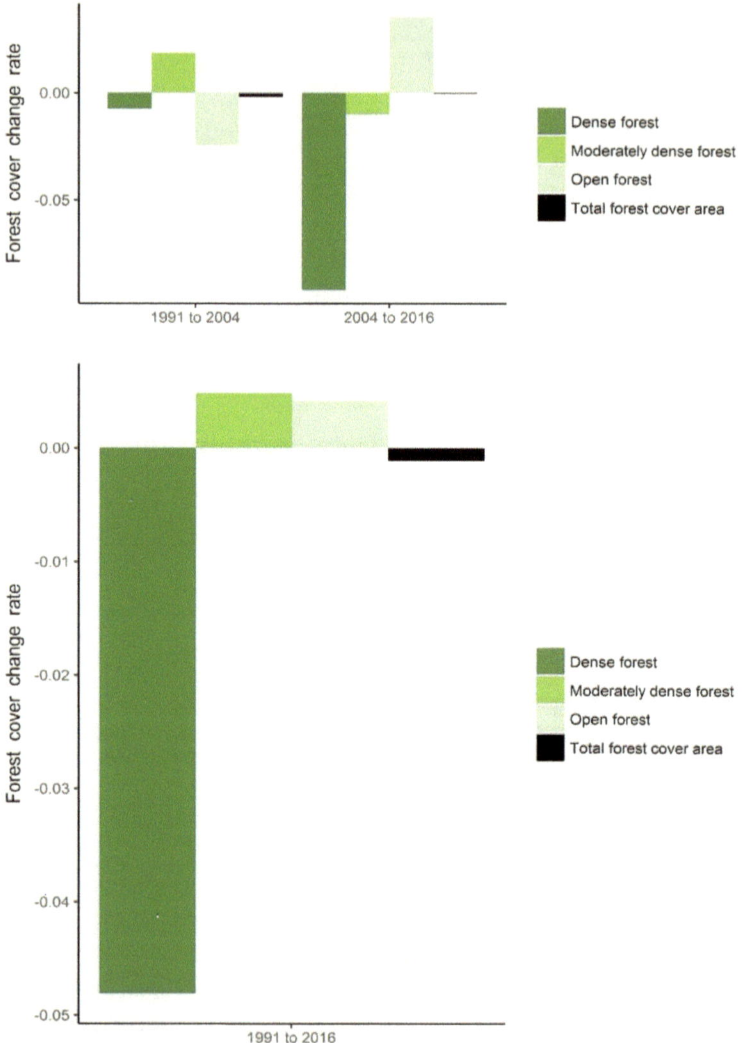

Figure 5: Comparative account of forest cover change rate of different forest types in and around Bailadila hill between 1991 to 2004, 2004 to 2016 and 1991 to 2016

3.3.2 Forest fragmentation

Dense forest: The area and landscape proportion was decreases from 1991 to 2004 and from 2004 to 2016. As number of patch and patch density were increased from 1991 to 2004 and then decreased from 2004 to 2016 in can be concluded that extent of forest fragmentation was increased from 1991 to 2004 and then decreased from 2004 to 2016. The largest patch size and largest patch index decreased from 1991 to 2004 and then increased from 2004 to 2016 which indicates that dominancy of a single, largest patch decreased from 1991 to 2004 and then increased from 2004 to 2016. It also indicates that extent of forest fragmentation was increased from 1991 to 2004 and then decreased from 2004 to 2016. The landscape shape become more irregular from 1991 to 2004 as edge length, edge density and landscape shape index increased and then become regular from 2004 to 2016 as edge length, edge density and landscape shape index decreased.

Moderately dense forest: The area and landscape proportion was increased from 1991 to 2004 and then decreased from 2004 to 2016. As number of patch and patch density were increased from 1991 to 2004 and then decreased from 2004 to 2016 in can be concluded that extent of forest fragmentation was increased from 1991 to 2004 and then decreased from 2004 to 2016. The largest patch size and largest patch index decreased from 1991 to 2004 and then again decreased from 2004 to 2016 which indicates that dominancy of a single, largest patch decreased from 1991 to 2004 and then again decreased from 2004 to 2016. It also indicates that extent of forest fragmentation was increased from 1991 to 2004 and then again increased from 2004 to 2016. The landscape shape become more irregular from 1991 to 2004 as edge length, edge density and landscape shape index increased and then become regular from 2004 to 2016 as edge length, edge density and landscape shape index decreased.

Open forest: The area and landscape proportion was decreases from 1991 to 2004 and then increased from 2004 to 2016. As number of patch and patch density were decreased from 1991 to 2004 and then increased from 2004 to 2016 in can be concluded that extent of forest fragmentation was decreased from 1991 to 2004 and then increased from 2004 to 2016. The largest patch size and largest patch index increased from 1991 to 2004 and then decreased from 2004 to 2016 which indicates that dominancy of a single, largest patch increased from 1991 to 2004 and then decreased from 2004 to 2016. It also indicates that extent of forest fragmentation was decreased from 1991 to 2004 and then increased from 2004 to 2016. The landscape shape become more regular from 1991 to 2004 as edge length, edge density and landscape shape index decreased and then become irregular from 2004 to 2016 as edge length, edge density and landscape shape index increased.

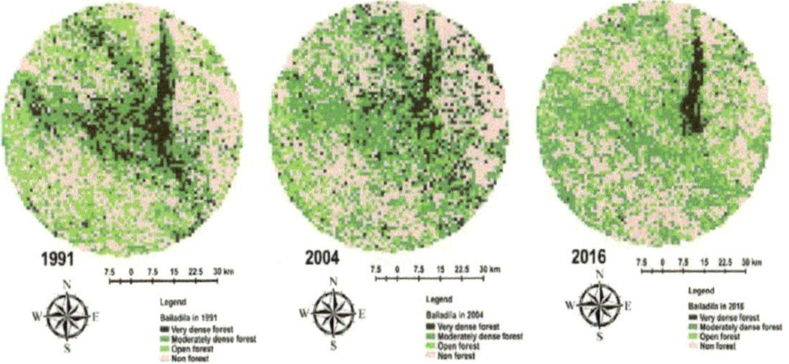

Plate 7: Forest cover fragmentation in and around Bailadila hill in 3 years - 1991, 2004 and 2016 *

* bigger depiction of this figure on the last page

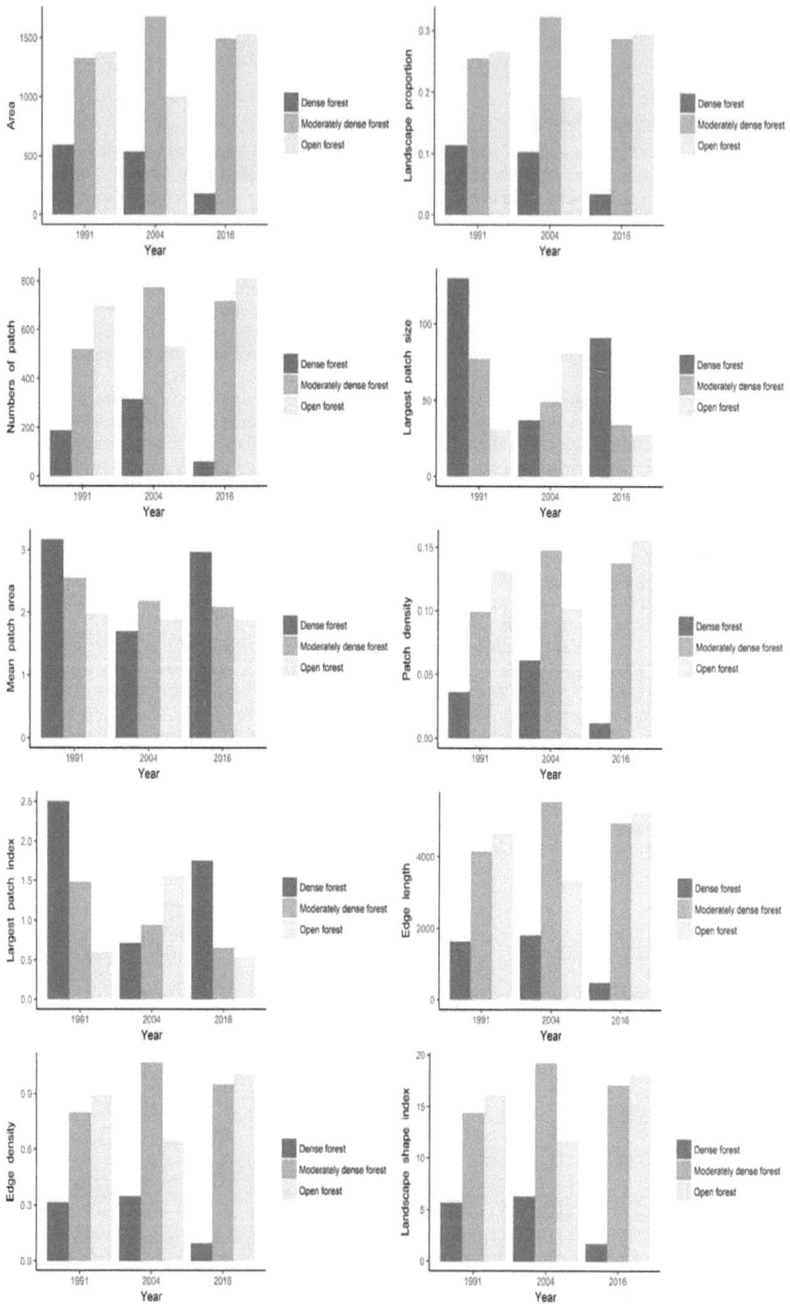

Figure 6: Comparative account of 10 forest fragmentation parameters of 3 forest cover types in and around Bailadila hill in 3 years - 1991, 2004 and 2016

Table 4: Forest cover change and fragmentation statistics in and around Bailadila hill in 3 years - 1991, 2004 and 2016 (areas are in in square km)

		1991	2004	2016
Total study area		5203	5203	5203
Total non forest area		1903	1979	1996
Total forest cover area		3300	3224	3207
% of total forest cover area		63.425%	61.964%	61.638%
Dense forest	Area	592	536	178
	Landscape proportion	0.114	0.103	0.034
	Number of patch	187	316	60
	Largest patch size	130	37	91
	Mean patch area	3.166	1.692	2.967
	Patch density	0.036	0.061	0.012
	Largest patch index	2.499	0.711	1.749
	Edge length	1638	1812	494
	Edge density	0.315	0.348	0.095
	Landscape shape index	5.677	6.28	1.712
Moderately dense forest	Area	1325	1682	1493
	Landscape proportion	0.255	0.323	0.287
	Number of patch	520	772	717
	Largest patch size	77	49	34
	Mean patch area	2.548	2.179	2.082
	Patch density	0.099	0.148	0.138
	Largest patch index	1.479	0.942	0.653
	Edge length	4156	5546	4950
	Edge density	0.799	1.066	0.951
	Landscape shape index	14.404	19.221	17.156
Open forest	Area	1383	1006	1536
	Landscape proportion	0.266	0.193	0.295
	Number of patch	698	533	811
	Largest patch size	31	81	28
	Mean patch area	1.981	1.887	1.894
	Patch density	0.132	0.102	0.156
	Largest patch index	0.596	1.557	0.538
	Edge length	4666	3366	5244
	Edge density	0.897	0.647	1.007
	Landscape shape index	16.172	11.666	18.175

3.4 Discussion

Over the years following publication of the theory of island biogeography (MacArthur & Wilson, 1963, 1967), the idea that patches of habitat are analogues of islands took root, becoming a central theme in conservation biology (Fahrig, 2013). Forest fragmentation generates forest patches and corridors which affects conservation aspects of wildlife, which in turn gave birth to famous SLOSS debate in 1970s. The acronym SLOSS stands for the phrase single large or several small and SLOSS debate refers to the question of whether it would be beneficial for the conservation of species to divide a given total amount of habitat into one large or several small habitat patches. Diamond (1975) proposed that a single large reserve is more preferable for biodiversity conservation than several smaller reserves whose total areas were equal to the large one. Simberloff and Abele (1976) challenged idea of Diamond (1975) and proposed that several small refuges may contain more species than a single large refuge. Cole (1981) supported the idea of Diamond (1975) and challenged idea of Simberloff and Abele (1976) and stated that larger refuges or islands generally will preserve more species than a series of small refuges of equivalent total area. By this way, conservation biologists were divided into two schools - "*school of single large refuge*" and "*school of several small refuges.*"

In recent times scientists are searching the questions of SLOSS debate using various ecological modelling, mathematical modelling and computer simulation techniques. The optimal size of habitat patches depends on the patch area scaling factor $z = z_{im} + z_{em} + z_{ex}$; where the three terms indicate how the key metapopulation processes, namely immigration, emigration, and extinction of local populations, scale with patch area (Ovaskainen, 2002). Habitat fragmentation affects different species in different ways (Laurance, 2010). Some species decline sharply or disappear in fragments, others remain roughly stable, and yet others increase, sometimes dramatically (Laurance, 2010). The most extinction- prone species are often found only in large reserves, favoring the single-large-reserve strategy, although small reserves scattered across a region can sustain certain locally endemic species that would otherwise remain unprotected (Laurance, 2010). Thus, the answer to SLOSS is, *"it depends"* (Laurance, 2010). So, as a whole, it can be inferred that as different faunal and floral taxa have different and unique ecological niche in the ecosystem, then effect of forest fragmentation, numbers and size of patch differ from taxon to taxon and these process cannot be generalized.

From the present study, it was found that the total forest cover area, especially dense forest cover area was reduced from 1991 to 2016 in the Bailadila hill range of southern Dandakaranya region. The forest fragmentation statistics shows that different forest fragmentation parameters such as landscape proportion, number of patch, largest patch size, mean patch area, patch density, largest patch index, edge length, edge density and landscape shape were not stable from 1991 to 2016 in the Bailadila hill range of southern Dandakaranya region. These unstable forest fragmentation statistics coupled with reduction of forest cover will definitely have impacts on the wildlife of the Bailadila hill range of southern Dandakaranya region. It is recommended that taxon specific (especially endemic, threatened and scheduled taxa) studies should be carried out by conservation biologists to assess the effects of forest fragmentation and reduction of forest cover on the wildlife of in and around Bailadila hill range of southern Dandakaranya region.

CHAPTER IV

General discussion and conclusion

Present study provides preliminary information about land use land cover changes and forest cover fragmentation in southern Dandakaranya region. The study finds that over the period of 25 years (1991 to 2016) the land use and land cover of the area changes, forest cover area, especially dense forest cover area is reduced and forest cover become more fragmented.

According to the 15th Indian Census (2011) the population density of southern Dandakaranya region is increasing. In Kanker district, population density was increased from 100/ square km in 2001 to 105/ square km in 2011, in Baster district, population density was increased from 119/ square km in 2001 to 135/ square km in 2011 and in Dantewada district, population density was increased from 53/ square km in 2001 to 64/ square km in 2011. This increased population density results in increased demand for lands for agriculture and settlements which results in loss and fragmentation of natural vegetation cover in southern Dandakaranya region. Beside, activities for mining, roadways and railways also add to the deforestation and forest fragmentation process.

Plate 8: Virgin forest of Kanger Valley, southern Dandakaranya on the bank of river Kolab

Beside mining and developmental activities, the shifting cultivation practice by local is another threat to this region because, due to shifting cultivation moist forests are converted rapidly into low quality open forest which leads to soil erosion and spread of alien invasive species such as *Lantana camara* and *Eupatorium adenophorum* (Govekar, 2008).

Dandakaranya landscape harbours one of the most ancient forest formations but today only a number of relict patches of vegetation representing interesting micro-habitats and special assemblages of biota exists their (Govekar, 2008). Dandakaranya region is the part of Global Frontier Forest (World's remaining large intact natural forest ecosystems which are relatively undisturbed and big enough to maintain all of their biodiversity, including viable populations of the wide-ranging species associated with each forest type) but it is under threatened category because ongoing or planned human activities (such as logging, agricultural clearing, and mining) will eventually degrade the ecosystem (Bryant et al., 1997).

From the present study, it can be concluded that the etymology of Dandakaranya has been changed. At Treta yuga, king Danda was cursed by Shukracharya because king Danda molested Shukracharya's virgin daughter Araja and due to cursed of Shukracharya entire kingdom of

Danda was converted to *"the forest of punishment"*. But now Dandakaranya can be called *"the forest that is being punished."* because virgin forest of Dandakaranya is being molested. Government, NGOs and people should join hands to protect Dandakaranya - an integral part of Indian culture, mythology, anthropogenic and biological diversity, otherwise in future the etymology of Dandakaranya will be again changed to *"the forest that gives punishment"* and in those days, curse of Shukracharya will not be confined in the Dandakaranya, but spread to the humansphere of the globe.

References

1. Bryant, D, Nielsen, D and Tangley, L. 1997. Last Frontier Forests. World Resources Institute, Washington D.C.

2. Central Pollution Control Board. 2007. Comprehensive Industry Document on Iron Ore Mining, Comprehensive Industry Document Series. Ministry of Environment and Forests, Govt. of India

3. Cole, B.J. 1981. Colonizing Abilities, Island Size, and the Number of Species on Archipelagoes. *The American Naturalist*, 117(5): 629-638

4. Diamond, J.M. 1975. The island dilemma: Lessons of modern biogeographic studies for the design of natural reserves. *Biological Conservation*, 7(2): 129-146

5. Diaz, S., Fargione, J., Chapin III, F.S. and Tilman, D. 2006. Biodiversity Loss Threatens Human Well-Being. *PLoS Biology*, 4:e277.

6. Didham, R.K. 2010. Ecological Consequences of Habitat Fragmentation. In: Jansson, R. (Ed.). Encyclopedia of Life Sciences. John Wiley & Sons Ltd.

7. Fahrig, L. 2013. Rethinking patch size and isolation effects: the habitat amount hypothesis. *Journal of Biogeography*, 40: 1649-1663

8. FAO. 2000. Land cover classification system (LCCS).

9. Govekar, R. 2008. Vegetation Characteristic and Special Habitats in the Transition Zone of Vidarbha – Dandakaranya, Deccan Plateau. pp. 155-161. In: Rawat, G.S. (Ed.). 2008. Special Habitats and Threatened Plants of India. ENVIS Bulletin: Wildlife and Protected Areas, Vol. 11(1). Wildlife Institute of India, Dehradun, India

10. Hilty, J.A., Lidicker Jr., W.Z. and Merenlender, A.M. 2006. Corridor Ecology: The Science and Practice of Linking Landscapes for Biodiversity Conservation. Island Press

11. Howard, N.K. 2005. Multiscale analysis of landscape data sets from northern Ghana: Wavelets and pattern metrics. ZEF - Ecology and Development Series No. 31. Centre for Development Research, University of Bonn

12. Indian Bureau of Mines. 2015. Indian Minerals Yearbook 2014, Part- III: Mineral Reviews (53rd Ed.) Iron Ore. Ministry of Mines, Government of India

13. Jaeger, J.A.G. 2000. Landscape division, splitting index, and effective mesh size: new measures of landscape fragmentation. *Landscape Ecology*, 15: 115-130

14. Jayanth, J., Kumar, T.A., Koliwad, S. and Krishnashastry, S. 2016. Identification of land cover changes in the coastal area of Dakshina Kannada district, South India during the year 2004-2008. *The Egyptian Journal of Remote Sensing and Space Sciences*, 19: 73-93.

15. Jung, M. 2013. LecoS - A QGIS plugin for automated landscape ecology analysis. *PeerJ PrePrints* 1:e116v2

16. Kumari, A. 2015. Runoff Water Management in Bailadila Range, Kirandul, India. Final Project Report submitted at Center for Research in Water Resources, University of Texas, Austin

17. Lambin, E.F., Geist, H.J. and Lepers, E. 2003. Dynamics of land-use and land-cover change in tropical regions. *Annual Review of Environment and Resources*, 28: 205-241

18. Laurance, W.F. 2010. Beyond Island Biogeography Theory: Understanding habitat fragmentation in the real world. In: Losos, J.B. and Ricklefs, R.E. (Ed.). The Theory of Island Biogeography Revisited. Princeton University Press.

19. MacArthur, R.H. and Wilson, E.O. 1963. An Equilibrium Theory of Insular Zoogeography. *Evolution*, 17(4): 373-387

20. MacArthur, R.H. and Wilson, E.O. 1967. The Theory of Island Biogeography. Princeton University Press.

21. McGarigal, K., Cushman, S.A. and Ene, E. 2012. FRAGSTATS v4: Spatial Pattern Analysis Program for Categorical and Continuous Maps. Computer software program produced by the authors at the University of Massachusetts, Amherst.

22. Ministry of Environment and Forests. 2007. Protection, Development, Maintenance and Research in Biosphere Reserves in India: Guidelines and Proformae. Government of India

23. Ovaskainen, O. 2002. Long-Term Persistence of Species and the SLOSS Problem. *Journal of Theoretical Biology*, 218: 419-433

24. Pattanaik, C., Reddy, C.S. and Reddy, P.M. 2011. Assessment of spatial and temporal dynamics of tropical forest cover: A case study in Malkangiri district of Orissa, India. *Journal of Geographical Sciences*, 21(1): 176-192

25. Puyravaud, J.P. 2003. Standardizing the evaluation of the annual rate of deforestation. *Forest Ecology and Management*, 177: 593-596.

26. Reddy, C.S., Jha, C.S., Dadhwal, V.K., Hari Krishna, P., Pasha, S.V., Satish, K.V., Dutta, K., Saranya, K.R.L., Rakesh, F., Rajashekar, G. and Diwakar, P.G. 2016. Quantification and monitoring of deforestation in India over eight decades (1930-2013). *Biodiversity and Conservation*, 25:93-116

27. Reddy, C.S., Rao, K.R.M., Pattanaik, C. and Joshi, P.K. 2009. Assessment of large-scale deforestation of Nawarangpur district, Orissa, India: a remote sensing based study. *Environmental Monitoring and Assessment*, 154:325-335

28. Samanta, K and Hazra, S. 2012. Landuse / Landcover change study of Jharkhali Island Sundarbans, West Bengal using Remote Sensing and GIS. *International Journal of Geomatics and Geosciences*, 3(2): 299-306

29. Sarkar, G., Corfu, F., Paul, D.K., McNaughton, N. J., Gupta, S.N. and Bishui, P.K. 1993. Early Archean crust in Bastar Craton, central India - geochemical and isotopic study. *Precambrian Research*, 62:127-137

30. Shannon, C.E. 1948. A mathematical theory of communication. *The Bell System Technical Journal*, 27: 379-423 and 623-56

31. Simberloff, D.S. and Abele, L.G. 1976. Island biogeography theory and conservation practice. *Science*, 91(4224): 285-286

32. Smith, T.M. and Smith, R.L.. 2012. Elements of ecology (8th Ed.). Pearson / Benjamin Cummings

33. Sridhar, M., Ramesh Babu, V., Chaturvedi, A.K. and Roy, M.K. 2015. Predictive GIS Modeling from Landsat, AGRS, Aeromagnetic and Ground Surveys for Uranium Exploration - A Case Study from Sonakhan Block, Chhattisgarh, India. *Journal of the Indian Society of Remote Sensing*, 43(2):347-362

34. Sudhakar, S., Pujar, G. and Anupama. 2015. Modern pollen analysis and ecological adaptations an parts of Jagdalpur forest division, Chattisgarh- an approach through RS & GIS techniques. *International Journal of Interdisciplinary Research in Science Society and Culture*, 1(2): 199-224

35. Thakur, T., Swamy, S.L. and Nain, A.S. 2014. Composition, structure and diversity characterization of dry tropical forest of Chhattisgarh using satellite data. *Journal of Forestry Research*, 25(4): 819-825

36. UNDP. 2005. Chhattisgarh Human Development Report - 2005. United Nations Organization

37. Zebardast, L., Yavari, A.R., Salehi, E. and Makhdoum, M. 2011. Application of Effective Mesh Size Metric for the Analysis of Forest Habitat Fragmentation inside the Defined Road Effect Zone of Golestan National Park. *Journal of Environmental Studies*, 37(580): 4-6

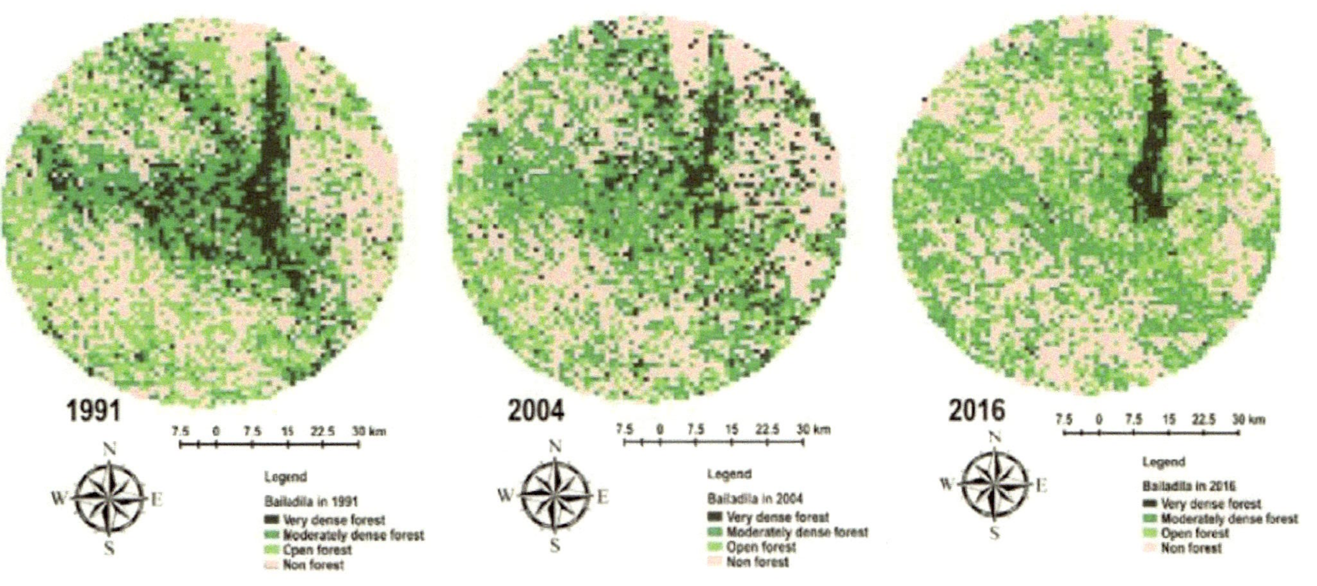
1991
7.5 0 7.5 15 22.5 30 km
N W E S
Legend
Bailadila in 1991
Very dense forest
Moderately dense forest
Open forest
Non forest
2004
7.5 0 7.5 15 22.5 30 km
N W E S
Legend
Bailadila in 2004
Very dense forest
Moderately dense forest
Open forest
Non forest
2016
7.5 0 7.5 15 22.5 30 km
N W E S
Legend
Bailadila in 2016
Very dense forest
Moderately dense forest
Open forest
Non forest

YOUR KNOWLEDGE HAS VALUE

- We will publish your bachelor's and
  master's thesis, essays and papers

- Your own eBook and book -
  sold worldwide in all relevant shops

- Earn money with each sale

Upload your text at www.GRIN.com
and publish for free